Astronomy: The secret behind black holes and what they truly are.

Brian C. Bivona

Table of contents

B. C. Bivona 4

INTRODUCTION

The Milky way

You should be able to see the Milky Way as a foggy ribbon across the sky if you travel somewhere dark, away from the strong city lights, and gaze up. The sky does resemble a glass of milk that has been spilled. The clouds are the combined light from thousands of stars, as you can see if you use a telescope to look at them more carefully.

Since the Milky Way is encasing us, we are seeing our galaxy edge-on from the inside. Grab a dinner plate and gaze at the edge of it so you can't see the galaxy's round form to get a better understanding. Only the plate's edge is visible.

Any bright, starry night will eventually see the Milky Way, a celestial belt, span across the sky.

Even though it has been known since antiquity, philosophers and scientists were only able to make educated guesses as to what it meant until relatively recently. With the development of the telescope, it was discovered that the Milky Way was the combined illumination of many dim stars.

English astronomer Thomas Wright proposed that this luminous band was exactly what one would expect to observe if the Sun were trapped in a flat disk of stars more than a century later. We now understand that the Milky Way is our galaxy's main edge-on structure. It took another two millennia to figure out further information, including the physical size of the galaxy.

Astronomers are still engaged in the process as they try to reconcile contradictory data and generate discoveries. Scientists must reconstruct the galaxy's structure while looking at it from within a disk where dust clouds dull and obstruct starlight, similar to charting a fog-bound city from a single junction.

In the 1920s, it became remarkably easier to understand the full magnitude of the Milky Way Galaxy and the cosmos as a whole. The discovery that "spiral nebulae" were whole galaxies like our own was made at that time thanks to a new generation of powerful telescopes and photography, which was referred to as "island universes" at the time.

Surveys revealed that young stars, gas, and dust were concentrated in the spiraling spiral arms of the majority of disk-shaped galaxies. Astronomers believed that our galaxy was also spiral-shaped. The first rough maps of the Milky Way's spiral arms were created in the 1950s by radio telescopes by observing the motion of gas clouds throughout the galaxy.

Surveys utilizing dust-penetrating infrared light have improved our understanding of the overall structure of our galaxy during the last 20 years. Two NASA spacecraft, the Wide-field Infrared Survey Explorer (WISE) and the Spitzer Space Telescope, as well as the ground-based Two Micron All-Sky Survey and Sloan Digital Sky Survey (SDSS), are among these initiatives. The central component of our galaxy, referred to as its "bulge," is a vast football-shaped star cloud that can be seen almost end.

These observations have also allowed astronomers to better define our galaxy's spiral arms, count star clusters and other phenomena in dust-obscured regions of the disk, and reveal this. The classification of the Milky Way was changed from spiral galaxy to barred spiral galaxy as a consequence of this finding.

To provide a full 3-D depiction of our planetary home, several ambitious and complementary initiatives are now underway. The 2013-launched Gaia mission from the European Space Agency is expected to provide location and motion data with unparalleled precision for almost a billion stars.

However, Gaia's primary coverage of optical wavelengths implies that the depth of the galaxy's disk that it can explore is constrained by intervening dust clouds. The Very Long Baseline Array (VLBA) is a facility that can measure distances and movements to a limited number of sources more precisely than Gaia since the dust has no impact on radio frequencies.

The VLBA has the highest resolving power of any telescope currently in use by connecting ten radio dishes that are spread out from Hawaii to St. Croix. Using this capacity, two studies, the Bar and Spiral Structure Legacy (BESSEL) survey and the VLBI Exploration of Radio Astrometry (VERA), are tracing the spiral structure of our galaxy by determining the positions and movements of areas where new stars are forming.

The outer edge of the Oort Cloud of comets, which is roughly 100,000 astronomical units (AU; the typical distance between the Earth and the Sun) distant or 1.6 light-years away, is where the galactic border is located. Here, the Sun's gravitational attraction is reduced to that of neighboring stars, and comets with orbits that take them this far from the Sun may completely escape the Sun's gravity. Although Proxima Centauri, at a distance of 4.22 light-years, is now the closest star, other stars have and will continue to fulfill this role in the past.

All stars revolve around the galactic center, although their orbits are more elliptical and inclined than those of the planets in our solar system. Moreover, one-third of the way into the disk and 27,200 light-years out from the Milky Way's core, the Sun is now located.

It is also located 90 light-years above the galaxy's midplane. The Sun's location oscillates above and below the galactic plane during each orbit, which takes around 240 million years to complete. The distribution and makeup of the stars around us progressively vary because the courses taken by the other stars in our neighborhood are somewhat different. Much closer stars than Proxima Centauri regularly pass the Sun.

For instance, researcher Ralf-Dieter Scholz of the Leibniz Institute for Astrophysics in Potsdam, Germany, found out in 2014 that a weak M dwarf star observed by WISE was around 20 light-years distant, making it a previously unidentified near neighbor. It was discovered by a team from the University of Rochester in New York, led by Eric Mamajek, that "Scholz's Star," which is a binary, moves rapidly away from us but not much else, indicating that it may have brushed the solar system.

The double, which now holds the record for the closest flyby of any known star, went far within the Oort Cloud, approaching 52,000 AU roughly 70,000 years ago. Any comets that are disturbed by this passage will take around 2 million years to enter planetary orbits, but the system's small mass just one-sixth that of the Sun and its trajectory through the outer Oort Cloud argue against any appreciable comet increase.

Dim, low-mass M dwarf stars like Proxima Centauri and Scholz's Star are typical of the Milky Way's stellar population. The vast majority of the 400 billion or so stars in the galaxy are probably M dwarfs, but since they don't radiate much visible light, we can still discover the ones that are near to our solar system using infrared surveys likeWISE. Mass is the fate of stars.

B. C. Bivona 13

Despite their low masses, M dwarfs can consume their nuclear fuel efficiently and continue to shine for billions of years after the Sun has died.

Some stars don't even glow very much. When they are young, they can create energy by fusing a rare kind of hydrogen called deuterium, but they never produce energy in their cores via real hydrogen fusion, the energy source that powers stars for the majority of their lifetimes. They are known as brown dwarfs and range in mass from 1.2 to 7% that of the Sun.

This class includes Scholz's Star partner. Brown dwarfs are minor stars that may be just as common as the main stars, with surface temperatures one-tenth that of the Sun. Within 16 light-years of the Sun, there are more than 50 known stars and brown dwarfs, yet only 10 of them are visible with the unaided eye.

Our perception of the stars in the galaxy is also distorted by the night sky. One-third of the top 100 stars in the sky are located within 100 light-years of one another. Among them is Sirius, the brightest star in the night sky, which is 8.6 light-years away, Procyon, 11 light-years away, Vega, 25 light-years away, Fomalhaut, 52 light-years away, Castor, 65 light-years away, Aldebaran, and Regulus (77). However, a further third of stars, such as Polaris (430), Antares (600), Betelgeuse (640), Rigel (860), and Deneb, are located more than 400 light-years distant (2,600).

These stars are tens of thousands of times more brilliant and have masses that are more than seven times that of the Sun. As a result, they use up their hydrogen fuel more quickly. These stars will finish their lives in magnificent supernova explosions far before the flares of our Sun die out.

Stars and clusters

Not merely because the most massive stars are the shortest-lived, but also because there are fewer and fewer stars as one moves up the mass scale. In thick, chilly molecular clouds, stars are formed. After a large star develops, the birth cloud begins to be dispersed by the star's intense ultraviolet radiation and a strong outflow known as a stellar wind, which reduces the number of nearby massive stars that may form.

Only a few Milky Way stars have energy outputs more than one million times that of the Sun. The two huge binary systems WR 25 and Eta Carinae, which shine with 6.3 million and 5 million solar luminosities, respectively, and are spaced around 7,500 light-years apart, are at the top of the list. The list also includes seven more stars in the Cygnus OB2 association and eight additional stars in the Carina Nebula.

B. C. Bivona 17

The spiral arms of our galaxy are mapped out in large part by massive stars. They are bright enough to be observed from a large distance, burst before leaving their star nurseries, and excite molecules like methanol and water in their evaporating birth clouds.

These molecules may transform into masers, the microwave equivalent of lasers, under the proper circumstances, which are typical in star-forming areas, and shoot amplified radio waves our way, producing beacons that cut through starlight-obstructing dust clouds.

Different kinds of star clusters are useful for tracing galactic structures. A few hundred light-years in size, OB associations are loose groups of 10 to hundreds of hot, young, massive, O- and B-type stars. The Scorpius-Centaurus OB2 association is the closest; it is located around 470 light-years distant and includes the red supergiant Antares. Approximately 15 million years ago, its oldest subgroup was formed, and about 5 million years ago, shock waves from a supernova helped start star formation in a nearby cloud.

Open star clusters are relatively close-knit groups of stars that originated in the same molecular cloud, such as the Hyades and Pleiades in Taurus (150 and 440 light-years apart, respectively), and the Beehive in Cancer (approximately 580 light-years apart).

These clusters will progressively spread over a few hundred million years and may include dozens to hundreds of stars in an area little wider than 50 light-years. There may be up to 100,000 stars in the Milky Way, but only approximately 1,200 have been cataloged.

Scientists are seeing the formation of new open clusters that are less than 2 million years old in regions like the Orion Nebula and the Eagle Nebula (1,350 and 7,000 light-years distant, respectively), where nascent stars have emerged from their birth clouds and set them aglow. Even before they leave their parent cloud, young clusters may be seen in the infrared.

The Federal University of Rio Grande does Sul in Brazil, under the direction of Denilson Camargo, published a paper in 2015 detailing the finding of hundreds of dust-covered clusters that were firmly entrenched in their host molecular clouds.

The Milky Way's disk contains OB associations, open clusters, and embedded clusters. However, globular clusters represent a fundamentally distinct class of cosmic structures. These clusters, which are essentially huge star balls, include tens of thousands to possibly a million stellar siblings in spheres that are fewer than 300 light-years in diameter. There are fewer than 200 in our galaxy, and they are all older than 10 billion years.

Globular clusters travel along radically inclined courses that carry them far above and below the disk as they circle the galaxy's core. However, more on that later. Scientists now know that the Milky Way has stolen at least some globular clusters.

Galaxy-wide structures

Early in the 20th century, astronomers used the distinctions between globular and open star clusters to get a general understanding of the Milky Way. Nearly the majority of the galaxy's gas and dust, which serve as the nucleus for young stars, are located inside the disk-shaped space in which open clusters revolve. The solar system is about a third of the way out in this disk, which is about 1,000 light-years thick and likely stretches 75,000 light-years from the galactic center.

Astronomy: Black holes

B. C. Bivona 22

The galactic bulge, which is around 12,000 light-years long and has the appearance of a football, is located in the middle of the disk and is made up mostly of ancient stars. We perceive the bulge obliquely, not too far from the end on, but its precise size, shape, and viewing angle remain rather obscure.

Up until recently, astronomers thought of the bulge as a kind of retirement community for aged stars, a population that quickly emerged when the galaxy came together via mergers with smaller galaxies around 10 billion years ago.

A broad variety of stellar ages, from 3 to 12 billion years, are seen closer to the midplane, although older stars do mostly populate portions of the bulge much above and below the disk, according to recent research. According to many lines of evidence, the growing disk's inherent instabilities had a significant role in the bulge population's formation.

The supermassive black hole in the heart of the bulge, around which everything else revolves, serves as the galaxy's anchor. It is estimated to be 4 million solar masses in mass. Regular observation of the galactic center reveals that it often flares in X-rays, which is a sign of stuff accelerating into a black hole.

However, this is nothing compared to what we know a monster black hole is capable of, and there is evidence that it has been more active in the past. Huge gamma-ray-producing bubbles that extended 25,000 light-years above and below the galactic center were discovered in 2010 by NASA's Fermi Gamma-ray Space Telescope. These bubbles are most likely the smoking gun of a tremendous outburst that occurred millions of years ago.

The number and placement of the disk's spiral arms, as well as its specific structural makeup, are still poorly understood. Recent radio studies of thousands of sources, including young star-lit nebulae, giant molecular clouds, water and methanol masers, stars in embedded clusters detected in the infrared, and stars in embedded clusters, seem to indicate that the Milky Way has four major spiral arms that start close to the galactic center and wind outward.

The Norma-Outer Arm, the Scutum-Centaurus Arm, and the Carina-Sagittarius Arm are in ascending sequence starting from the center and advancing toward the Sun. The Perseus Arm and the outer arc of the Norma-Outer Arm are located farther out.

For a very long time, astronomers believed that the solar system was situated in a starry spur towards the inner border of the Perseus Arm. Yet our "spur" is a substantial structure with as much massive star creation as the nearby main arms, which is one of the first startling findings from the BESSEL and VERA investigations. Astronomers aren't yet certain whether to label our neighborhood of the galaxy as a branch of the Perseus Arm or a separate piece.

The disk also has additional surprises in store. The size of the SDSS dataset has increased by nearly 50% from prior estimates according to 2015 research performed by Yan Xu at the Chinese Academy of Sciences in Beijing. Around 50,000 light-years from the disk's core, the number of stars had looked to decline, but SDSS discovered what appeared to be a massive ring of stars 10,000 light-years further out.

According to the new research, there are at least four ripples in the disk above and below the galactic plane that are moving stars, creating an illusion that causes this. The disk, which begins around 6,500 light-years from the Sun and extends at least 50,000 light-years distant, is agitated as we stare out of the galaxy from the solar system. It moves up a few hundred light-years, then down, then up, and then down again. Undiscovered waves could still be present.

The vibrations might have been caused by tiny galaxies around our own. One, in particular, the Sagittarius Dwarf Spheroidal, is orbiting the Milky Way and has gone through the disk many times. As it does so, it is progressively disintegrating into streams of stars. The situation was like a stone thrown into quiet water as the gravitational effect of a satellite galaxy plunging through the disk could produce ripples.

The spiral structure may be influenced by satellite galaxies ripping across the disk, according to simulations. It's also remarkable how closely the newly discovered ripples match the spiral arms of the Milky Way.

The galactic halo, which is a sphere controlled by globular clusters and satellite galaxies and where stars have been stripped from them, is where the disk is located. Most galaxies, including our galaxy, may have formed by consuming several smaller galaxies. Today, we see star streams connected to several tiny satellites, and it appears that the Milky Way has stolen numerous globular clusters from the Sagittarius Dwarf Spheroidal.

Omega Centauri, the biggest and brightest globular cluster, is around 17,000 light-years distant and has a more intricate star structure than others. It may be the remnant bulge of a dwarf galaxy that our galaxy destroyed long ago, according to researchers.

But much of the Milky Way's bulk is still hidden from view. A gravitational impact that extends far beyond the structures we can see may be seen in the movements of stars orbiting our galaxy and other galaxies. According to studies, the

Milky Way is surrounded by a roughly spherical halo of dark matter that spans 900,000 light-years, or nearly six times the disk's diameter.This substance, which accounts for about 27% of the universe, produced the gravitational framework that guided ordinary matter towards the formation of galaxies like our own.

Major new insights have already been gained from the present phase of cosmic research, but many more questions remain. We will finally be able to perceive our island universe in the same way we view other galaxies: as a whole cosmic entity, a whole larger than the sum of its parts when astronomers combine the findings of this study over the next ten years.

SECTION 2

Evolution of Black Holes

Don't be misled by the term; a black hole is everything but black. Instead, there is a lot of stuff crammed into a little space; imagine a star 10 times as big as the Sun being crammed into a sphere the size of New York City. As a consequence, there is a gravitational force that is so powerful that even light cannot escape. These bizarre objects, which are often regarded as the most interesting things in space, have just been given a fresh perspective by NASA instrumentation.

Since ancient times, there has been speculation about the possibility of a vast, dense object in space from which light could not escape. Einstein's theory of general relativity, which demonstrated when a big star dies, leaving behind a tiny, dense remnant core, is most renowned for foretelling black holes. The calculations demonstrated that if the core's mass is more than about three times that of the Sun, the force of gravity will outweigh all other forces and result in the creation of a black hole. Black holes are invisible to telescopes that look for light, x-rays, or other types of electromagnetic radiation.

But by seeing how they affect neighboring matter, we may deduce the existence of black holes and study them. A black hole will accrete matter, or pull it inward, if it passes through a cloud of interstellar matter, for instance. A normal star may experience a similar behavior as it approaches a black hole. In this situation, when the star is being drawn into the black hole, it may rip apart. The heated and accelerated attracting matter radiates x-rays into space as it accelerates and warms up.

B. C. Bivona 32

Recent findings provide some fascinating evidence that black holes have a profound impact on the regions surrounding them. They release strong gamma-ray bursts, eat neighboring stars, and stimulate or hinder the birth of new stars depending on where they are located.

Black holes may number over 100 million in the Milky Way, yet it is incredibly challenging to find these ravenous monsters. Sagittarius A*, a supermassive black hole, is located in the center of the Milky Way. According to a NASA statement, the enormous structure is located 26,000 light-years from Earth and has a mass that is around 4 million times that of the sun.

The Event Horizon Telescope (EHT) team obtained the first picture of a black hole in 2019. Scientists from all around the globe were enthralled by the stunning image of the black hole at the heart of the M87 galaxy, located 55 million light-years from Earth.

In 1916, Albert Einstein's general theory of relativity made the first prediction about the existence of black holes. Many years later, in 1967, American astronomer John Wheeler first used the phrase "black hole." Until recently, black holes were solely understood as hypothetical entities.

Cygnus X-1, in the Milky Way's constellation of Cygnus, the Swan, was the first black hole ever found. According to NASA, the first indications of the black hole were discovered in 1964 when a sounding rocket discovered astronomical X-ray sources (opens in new tab). Astronomers discovered in 1971 that the source of the X-rays was a brilliant blue star circling a mysterious black object. It was hypothesized that the discovered X-rays were caused by stellar material being "gobbled" up by the dark object—an all-consuming black hole—after being peeled away from the blazing star.

The Space Telescope Science Institute (STScI) estimates that one out of every thousand stars have enough mass to develop into a black hole. Since there are more than 100 billion stars in the Milky Way, there must be 100 million black holes in our galaxy. Though it may be challenging to find black holes, NASA estimates(opens in new tab) that the Milky Way may contain up to a billion-star black hole.

The "Unicorn" black hole, which is 1,500 light-years from Earth, is the nearest one to our planet. The moniker has two distinct meanings. The black hole candidate is almost unique due to its very low mass, which is around three times that of the sun, as well as the fact that it is located in the constellation Monoceros ("the unicorn").

The Event Horizon Telescope (EHT) team published the first black hole picture ever taken in 2019. While the telescope was studying the event horizon or the region beyond which nothing can escape from a black hole, the EHT discovered the black hole at the heart of galaxy M87. The picture depicts the abrupt loss of photons (particles of light). Additionally, since astronomers are aware of what a black hole looks like, it opens up a whole new field of study for them.

A fresh image of the massive structure in the heart of M87, as seen in polarized light, was released by scientists in 2021. The new picture reveals significantly more information about the black hole because polarized light waves vary in direction and brightness from unpolarized light waves. The picture makes it obvious that the ring of the black hole is magnetized since polarization is an indication of magnetic fields.

How do black holes appear?

The singularity, the outer and inner event horizons, and black holes all have three "layers."

A black hole's event horizon is the area around its mouth beyond which light cannot go. A particle cannot depart the event horizon once it has done so. Across the event horizon, gravity remains constant.

The singularity, or one point in space-time where the mass of the black hole is concentrated, is the area within a black hole where the object's mass is located.

Black holes are invisible to scientists, unlike stars and other celestial phenomena. As gas and dust are sucked into the massive objects, black holes must release radiation that astronomers must instead detect. However, because of the dense gas and dust that surround them, supermassive black holes that are in the galactic center may get obscured, obscuring their telltale emissions.

Sometimes, instead of being pulled into the black hole's mouth as the matter is attracted toward it, matter ricochets off the event horizon and is flung forth. It produces bright jets of matter moving at almost relativistic speeds. These strong jets may be seen from a vast distance, even though the black hole itself is not visible.

The EHT's photograph of a black hole in M87 (published in 2019) required two years of investigation even after the photos were obtained, which was an exceptional effort. This is so that the incredible quantity of data produced by the partnership of telescopes, which spans several observatories globally, can't be sent via the internet.

With time, scientists hope to photograph other black holes and compile a collection of images of the objects. Sagittarius A*, the black hole at the heart of our own Milky Way galaxy, is most likely the next target. A 2019 research found that Sagittarius A* is fascinating because it is quieter than predicted, which may be because magnetic fields are suppressing its activity. Another investigation that year revealed that Sagittarius A* is surrounded by a cold gas halo, providing new information about the surroundings of black holes.

Types of Black holes

Stellar black holes, supermassive black holes, and binary black holes are the three categories of black holes that astronomers have so far identified.

Stellar black holes (Small but mighty)
A star may collapse or fall into itself after it has used up all of its fuel. The new core of smaller stars (those with masses up to around three times that of the sun) will eventually transform into a neutron star or white dwarf. But as a bigger star disintegrates, it keeps becoming smaller and smaller until it becomes a stellar black hole.

Black holes created are when individual stars collide are comparatively tiny but immensely dense. One of these objects has a diameter of a city and is more than three times the mass of the sun. As a result, the gravitational attraction on things nearby is very strong. The gas and dust from the surrounding galaxies are subsequently ingested by stellar black holes, which stop them from contracting in size.

Supermassive black holes: where giants are created.
Although there are many little black holes in the cosmos, supermassive black holes predominate. Although they are millions or even billions of times more massive than the sun, the diameter of these giant black holes is around the same. It is hypothesized that such black holes, including the Milky Way, reside at the heart of almost every galaxy.

The origin of such massive black holes remains a mystery to scientists. Once these giants have formed, they continue to increase in size by absorbing mass from the gas and dust that surround them, which is abundant at the core of galaxies. Hundreds of thousands of microscopic black holes may combine to become supermassive black holes.

Large gas clouds that collide and quickly accrete mass might be to blame. A stellar cluster collapsing, or a collection of stars collapsing at once is a third possibility. Fourth, vast collections of dark matter may give birth to supermassive black holes. We can directly detect this material by its gravitational pull on other things, but we are unable to directly view the dark matter, thus we do not know what it is made of.

Binary form black holes: Twice the problem
Astronomers discovered gravitational waves from merging star black holes in 2015 using the Laser Interferometer Gravitational-Wave Observatory (LIGO).

Two ideas exist about the formation of binary black holes. According to the first theory, two stars that were born together and perished violently at around the same time gave rise to two black holes that formed in a binary at or near that same moment. The two black holes that were left behind would have had the same spin orientation as the partner stars.

According to the second hypothesis, two black holes in a star cluster merge at the cluster's core. According to the LIGO Scientific Collaboration, these partners would have unpredictable spin orientations when compared to one another. This creation scenario is supported by LIGO's discoveries of partner black holes with various spin orientations.

Facts about black holes

- Theoretically, if you went into a black hole, you would be stretched out like spaghetti by gravity, but you would die before you reached the singularity. But according to 2012 research in the journal Nature, the event horizon will behave like a wall of fire due to quantum phenomena, instantaneously killing you.

- Black holes are not bad. The big black hole is most clearly not a vacuum, which is what causes suction. Instead, like with everything that has gravity acting on it, like the Earth, items fall into them.

- Cygnus X-1 is the first object that has been identified as a black hole. In a casual bet made in 1974 between Stephen Hawking and scientist Kip Thorne, Hawking wagered that the source of Cygnus X-1 was not a black hole. 1990 saw Hawking admit defeat.

- The Big Bang may have been followed quickly by the formation of tiny black holes. Some areas may have been compressed into tiny, dense black holes with masses smaller than the sun by the rapidly expanding space.

- A star may be shattered if it comes too near to a black hole.
- The Milky Way is thought to contain between 10 million and 1 billion star black holes, each with a mass approximately equivalent to three suns.
- Black holes are still a great subject for science fiction literature and film. Check out "Interstellar," which significantly included science thanks to Thorne. Scientists now have a better idea of how far away stars could seem when viewed close to a rapidly rotating black hole because of Thorne's work with the film's special effects crew.

Most common inquiries:

1.The size of a black hole

Black holes appear in a variety of sizes, and the amount of matter they contain determines how big they are (their mass). Some are the leftovers of a massive star that disintegrated. For a star to turn into a black hole, it must be substantially more massive than our Sun.

These dark holes are only a few kilometers in diameter. The cores of certain galaxies have also been found to contain black holes. These enormous black holes are made up of the equivalent of 100 million or more suns' worth of stuff. These black holes have a diameter of several million kilometers.

2.Exactly how are black holes found?

Black holes are discovered when nearby material, like gas, is drawn into a disk by the gravity of the black hole. Because of how quickly the gas molecules in the disk spin around the black hole, X-rays are produced as they heat up. From Earth, these X-rays may be seen. By keeping an eye out for changes in the velocity of nearby stars, black holes may also be found.

3.Is our galaxy's core home to a black hole?
Yes, our galaxy's core is home to a huge black hole. At a distance of around 24,000 light-years from Earth, it has a mass of three million suns. The centers of the majority of massive galaxies are considered to naturally contain enormous black holes, and many of them have already been found.

There is just too much distance between Earth and the black hole at the core of our galaxy for there to be any threat.

4.Can the Sun turn into a black hole?
The Sun is much too tiny to ever turn into a black hole, however. Before a star disintegrates into a black hole, it has to be much more massive than the Sun.

5.Why do black holes form?

According to scientists, areas where stuff becomes very dense are where black holes are produced (where a huge amount of material is crammed into an extremely small space). This may take place at the hubs of massive galaxies or whenever a gigantic star contracts and shrinks as it nears the end of its existence. A black hole forms when a patch of substance becomes so dense that light cannot escape from it.

Conclusion

Why black holes have such a terrifying reputation is simple to understand. They aren't exactly the best place to spend your vacation time.

You may see a disk of incandescent material around a black hole if you were to approach it. It is possible for gas and other materials to be drawn into orbit around a black hole. Friction is produced as a result, which generates light and heat. This is known as an accretion disk.

An accretion disk may become as hot as millions of degrees Celsius. It emits a lot of high-energy X-rays as a consequence. This aids in the discovery of black holes, but it also implies that being too close to one might be harmful to your health.

Things would grow much stranger if you managed to survive the accretion disk's heat and radiation. There is a region dubbed the "photon sphere" that lies just beyond the event horizon of certain black holes.

Gravity is powerful enough in the photon sphere to bend light around the black hole. Theoretically, light might bounce off your head's back, travel around a black hole, and then strike your eyes. You would be looking at the back of your own head as you passed the event horizon!

Things would get quite gloomy after you were over the event horizon. The gravitational force on your feet would be far greater than the gravitational pull on your head if you were falling toward the center of the black hole feet first. You would become like spaghetti, stretched and compressed. Scientists have given this phenomenon the moniker "spaghettification."

The whole thing may seem frightening, but it's cool, really.

Black holes pose absolutely no threat to us. Similar to tigers, it's not a good idea to put your head in one of their mouths, but chances are you won't run across one on your walk to the store.

Black holes don't hunt as tigers do. They aren't roving the universe devouring planets and stars. There is no possibility of Earth ever falling into a black hole since there are no black holes in the vicinity of our Solar System.

In actuality, the distance between Earth and the nearest black hole is 1500 light years. If we were to use a rocket to go there, it would take 30 million years.